# SCIENCE ON THE EDGE

# CLONING

● ● ● ●

## WRITTEN BY
## DON NARDO

**BLACKBIRCH®**
PRESS

**THOMSON**
———✦———
**GALE**

San Diego • Detroit • New York • San Francisco • Cleveland • New Haven, Conn. • Waterville, Maine • London • Munich

# THOMSON
✳ ™
## GALE

*For more information, contact*
The Gale Group, Inc.
27500 Drake Rd.
Farmington Hills, MI 48331-3535
Or you can visit our Internet site at http://www.gale.com

Photo credits: cover, pages 9, 14, 15 © Getty Images; page 4 © Courtesy of United States Holocaust Memorial Museum; pages 5, 6, 7, 8, 10, 11, 12, 13, 16, 18, 20, 21, 22, 23, 24, 25, 32, 33, 34, 36, 37, 38,40, 41, 42, 43 © CORBIS; pages 17, 19, 20, 26, 29, 30, 31, 39 © Corel Corporation; page 44 © AP Wide World

### LIBRARY OF CONGRESS CATALOGING-IN-PUBLICATION DATA

Nardo, Don, 1947-
  Cloning / by Don Nardo.
    p. cm. — (Science on the edge series)
Summary: Discusses the history of the concept of cloning and the pros and cons of cloning animals and humans.
Includes index.
  ISBN 1-56711-782-1
  1. Cloning—Juvenile literature. [1. Cloning.] I. Title. II. Series.
QH442.2 .N37 2003
660.6'5—dc21                                                                        2002010369

Printed in China
10 9 8 7 6 5 4 3 2 1

TABLE OF CONTENTS

Introduction: *Drama Surrounds Cloning* . . . . . . . . . . . . . . . . . .4

Chapter One: *Cloning Is Nothing New* . . . . . . . . . . . . . . . . . .7

Chapter Two: *The Potential Benefits of Animal Cloning* . . . . . .22

Chapter Three: *The Promise of Human Cloning* . . . . . . . . . . . .33

Glossary . . . . . . . . . . . . . . . . . . . . . . . . . . . . . . . . . .45

For Further Information . . . . . . . . . . . . . . . . . . . . . . . . . .45

About the Author . . . . . . . . . . . . . . . . . . . . . . . . . . . . .47

Index . . . . . . . . . . . . . . . . . . . . . . . . . . . . . . . . . . .47

## DRAMA SURROUNDS CLONING

Cloning, especially human cloning, is one of today's most controversial issues. Many people strongly believe that humans should not be cloned. At least to some degree, such negative views are based on false ideas of what cloning really is and what it can do.

Do you think cloning is an evil technology that will enable a small group of scientists to create a master race that will take over the world? If so, it is probably because popular books and movies have misled people about the realities of cloning. About 40 years ago, science fiction writers and filmmakers began to deal with the concept of human cloning. They were—and still are—fascinated by tales about mad or evil scientists who threaten to destroy humanity. So they added cloning to a list of potentially dangerous technologies, such as robots, atomic energy, and artificial intelligence. Therefore, it is not surprising that many people began to believe that human cloning would probably be used in an unethical manner.

Two movies in particular helped shape these negative attitudes. In 1973, actor/director Woody Allen released the film *Sleeper.* In this spoof, someone in a future society keeps a dead dictator's nose alive for almost a year. The person then tries to clone a copy of the dead man from the nose.

**Many fear cloning will be used to bring back people like Nazi dictator Adolf Hitler.**

Scientists have been cloning plants for years.

In a more serious and disturbing vein, in 1976, novelist Ira Levin published *The Boys from Brazil.* It is the chilling story of a Nazi doctor who manages to create 94 clones of the infamous dictator Adolf Hitler from cells Hitler contributed before his death at the end of World War II. The doctor's plan is to end up with at least one new Hitler, who will grow up to fulfill the Nazi dream of world conquest. In 1978, a popular film version of the novel was released.

Until recently, most people outside of the scientific community did not realize that such depictions of cloning are purely fanciful. It was also not widely known that cloning—both natural and artificial—has gone on for millions of years. Many plants and animals reproduce by cloning. Humans have also cloned plants for agriculture and industry. Whether human cloning will ever occur remains to be seen. If it does, though, it will almost certainly not be used to resurrect dictators and create evil armies. Instead, scientists foresee that both animal and human cloning will be used to wipe out disease and to end other forms of suffering.

Clones are created from the cells of existing plants or animals.

## CHAPTER 1

## CLONING IS NOTHING NEW

Put simply, cloning is the creation of a new plant or animal from the cells of an existing one, so that the offspring is genetically—and therefore physically—identical to the parent. Genes are the tiny particles in the cells of all living things that carry the blueprints for reproduction. Specific genes determine physical traits such as eye and hair color and body size and type. In sexual reproduction, a male and female mate and their genes mix. As a result, the offspring has genes from both the mother and the father. A clone, by contrast, is the offspring of any living thing that uses asexual means to reproduce. In asexual reproduction, a cell, plant, or animal has offspring on its own, without sexual union or aid from another living thing. In this process, the parent passes on its DNA (the most important chemical that makes up the genetic material) to the offspring. When the offspring grows up, it is a genetically identical copy of the parent.

Yeast cells reproduce asexually.

Many people are surprised to learn that human beings did not invent cloning. For millions of years, cloning has been a natural process that has helped both plants and animals to reproduce. One-celled creatures, including bacteria, use cloning to make copies of themselves. The same is true of various kinds of multi-celled plants and animals.

When human beings came along, they learned to grow crops to feed themselves. Over time, they began to imitate nature by cloning various varieties of fruits and vegetables. People realized that they could take the biggest, tastiest, most disease-free kind of a certain fruit, then clone it to produce larger and more valuable harvests. Later, in the twentieth century, the use of cloning to grow food became a large-scale business in the United States and other developed countries.

**Garlic plants reproduce by cloning themselves.**

Scientists have used cloning to produce large, tasty, disease-free apples and other fruits and vegetables.

The next logical step came when some scientists decided to apply the same idea to livestock. They knew that if they could find a way to clone animals such as cattle, pigs, and sheep, they could produce better herds. As it turned out, however, it was much more difficult to clone animals than it was to clone plants. In fact, the first major success in the cloning of livestock did not occur until 1996.

**Scientists hope to create hardier livestock by cloning. Dolly the sheep was successfully cloned in 1996.**

# AN EVIL DICTATOR RESURRECTED?

The world breathed a sigh of relief when notorious German dictator Adolf Hitler committed suicide toward the end of World War II. In the 1970s, though, a popular novel gave many people nightmares about what might happen if Hitler could somehow be resurrected through cloning.

Nazi doctor Josef Mengele's plot to clone Hitler was the subject of the movie *The Boys from Brazil.*

Many people's views about cloning were shaped by this scary novel—*The Boys from Brazil* by Ira Levin. In Levin's frightening tale, Nazi doctor and war criminal Josef Mengele (who was a real person) manages to create 94 clones of Hitler from cells donated by Hitler before his death. The cloned boys, who are all physically identical, live in various parts of the world with families specifically chosen because they resemble the family in which Hitler himself grew up. Mengele does this because he knows that home environment will play a crucial role in the creation of a new Hitler. The act of copying the genes through cloning duplicates only the physical characteristics of the deceased person. In theory, to duplicate a person's personality, it is necessary to create an environment in which the clone will have experiences similar to those of the donor. Mengele does not live long enough to find out if this theory is valid. A Jewish Nazi-hunter finds out about the evil plan, wrecks it, and kills Mengele.

# CLONING IN NATURE

Long before humans learned to clone plants and animals, nature was a very good cloner. The earth's first and smallest life forms were one-celled organisms that appeared in the oceans more than a billion years ago. These organisms reproduced by cloning.

Many of nature's simplest and most numerous creatures still reproduce in this way. Among them are all forms of bacteria and most kinds of protozoa. Blue-green algae and some yeasts also clone themselves to reproduce. Every one of these creatures in all parts of the world is a clone that came from an earlier clone that came from an even earlier clone. If it were possible to trace these generations of clones backward, the line of descent would reach back to the very first plants and animals that lived in Earth's earliest seas.

Some more highly developed kinds of plants and animals reproduce by cloning, too. One example is the water hyacinth. This plant puts out thin stems, and each stem grows into a genetic copy of the parent plant. After that, the new hyacinths in turn clone themselves, and the process continues to repeat.

**Blue-green algae has cloned itself for millions of years.**

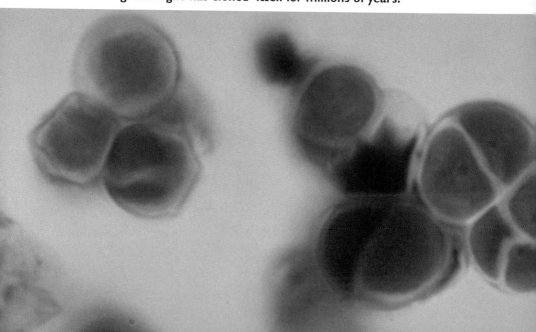

Water hyacinths reproduce by cloning.

# CLONING IN AGRICULTURE

At least 4,000 years ago, and possibly earlier, gardeners and farmers began to clone plants. A common method was to take a twig or cutting from a plant and put it in a pot filled with rich soil. Soon, roots began to sprout, which signaled that it was time to move the twig or cutting to the ground. There, the twig grew into a new version of the old plant. Farmers learned that if they took cuttings from the healthiest plants, they were able to produce larger numbers of new, healthy plants. These farmers did not know about genes and DNA, so they did not understand the scientific principle behind what they did. All that mattered to them was that the process brought good results.

**Some grapes grown in Europe are clones of ancient Roman grapes.**

The results of some of these ancient cloning experiments were not only beneficial, but also long-lasting. In Italy and other parts of Europe today, some varieties of grapes are clones of grapes that Roman farmers grew more than 2,000 years ago. Over many centuries, Italian vineyards continued to clone these same vines. In the process, the vines' unique traits were passed down through the generations.

Many of today's fruits and vegetables are also clones of versions first grown long ago. A group of apples in a supermarket will all be roughly the same size, shape, and color. This did not occur simply because the farmer diligently fertilized and watered the apple trees. The apples are similar because they are clones. Some of the apple varieties that are popular today were developed in the early twentieth century. A few, however, originated much earlier. The modern trees that produce the Cox's Orange Pippin apple are all clones of one apple tree that was planted about 200 years ago.

Most varieties of apples grown today are clones.

There are both advantages and disadvantages to the use of cloning techniques to grow food. Among the advantages are crops that look more appealing, taste better, and are more healthful for people to eat. Cloning also uses time and money more efficiently. Farmers do not have to waste time growing many versions of the same plant in the hope that a few will turn out to be healthy and valuable.

So far, farmers have found only one major disadvantage of cloning foods. A new plant that is genetically the same as an older plant faces an equal danger that it might get the same diseases as the older one. That means that if a disease can kill a certain fruit or vegetable, it could also infect and destroy most or even all of its clones. Techniques to reduce the risk of such a disaster rely on the fact that most diseases attack only one or a fairly small number of plant or animal species. First, two or more somewhat different kinds of a crop are grown. Then each kind is cloned. This gives the crop enough diversity so that a disease that attacks one cloned line will most likely not destroy the plants from the other lines.

# EARLY ATTEMPTS TO CLONE ANIMALS

Farmers could continue to profit from a prize-winning cow by cloning it.

After years of cloning plants, some scientists realized that it might be just as profitable to clone livestock as it was to clone fruits and vegetables. With cloning, farmers could select their best animals and breed whole herds of them. For example, a farmer could be sure that calves cloned from an excellent milking cow would all grow up to be great milking cows, too.

Animal cloning also held promise for laboratory scientists who need mice, rats, and other animals for experiments. Usually, scientists want all the lab animals used in an experiment to be as similar as possible. This way, the results will not be affected or confused by any physical or other differences in the animals.

For these reasons, the cloning of animals seemed like a great idea. The trouble was that the first scientists who tried it found that it was much more difficult to clone animals than plants. In the 1930s, a Nobel Prize-winning biologist named Hans Spemann came up with a possible way to clone an animal. The technique would be to remove an egg from a female animal and take away the genetic material from the egg's nucleus, or center. Then the genetic material from a cell of the animal to be cloned would be placed inside the nucleus. Soon, the egg would grow into an embryo that could be implanted into the womb of a female of the species to be cloned. The result, Spemann predicted, was that a cloned offspring would later be born.

Spemann wanted to try this technique, so he did some simple experiments. His efforts were unsuccessful, however, and he had to give up. Spemann was on the right track; he failed mainly because he needed tiny instruments, as well as a deeper knowledge of genetics, that were not yet available. In the 1950s, a few scientists with slightly better equipment and information had some minor

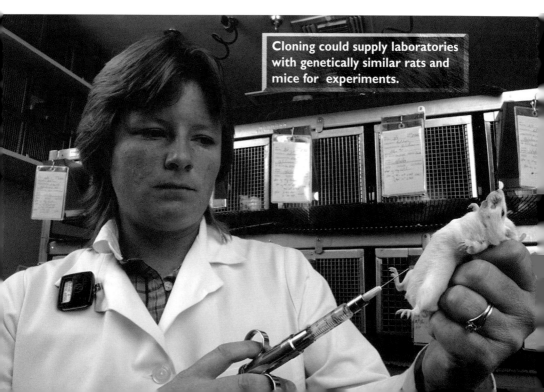

Cloning could supply laboratories with genetically similar rats and mice for experiments.

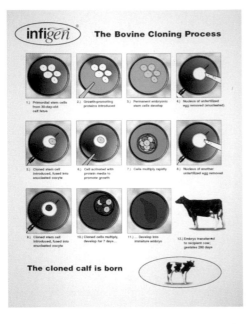

infigen® The Bovine Cloning Process

1.) Primordial stem cells from 30-day-old calf fetus
2.) Growth-promoting proteins introduced
3.) Permanent embryonic stem cells develop
4.) Nucleus of unfertilized egg removed (enucleated)

5.) Cloned stem cell introduced, fused into enucleated oocyte
6.) Cell activated with protein media to promote growth
7.) Cells multiply rapidly
8.) Nucleus of another unfertilized egg removed

9.) Cloned stem cell introduced, fused into enucleated oocyte
10.) Cloned cells multiply, develop for 7 days...
11.) ... Develop into immature embryo
12.) Embryo transferred to recipient cow; gestates 280 days

**The cloned calf is born**

success. They took the nucleus out of a frog's egg and replaced it with the nucleus from the cell of a frog embryo. The cloned frog embryos they made did not live long, however.

Similar experiments were done over the years that followed, most often with the same discouraging results. It was not until 1996 that a major breakthrough occurred. On July 5 of that year, a sheep that had been cloned from a cell of a living adult sheep was born at the Roslin Institute in Roslin, Scotland. The team of researchers who created the clone, headed by embryologist Ian Wilmut, named the new arrival Dolly.

The experiments that produced Dolly first began in 1986. The researchers' goal was not simply to be the first lab in the world to clone an animal. The Roslin Institute wanted to find better ways to use sheep to make drugs that might help fight human diseases. Earlier scientists had found a way to engineer sheep genetically so that their milk contained a drug called alpha-1 antitrypsin (AAT). This drug showed a lot of promise for the treatment of cystic fibrosis and emphysema,

**The chart above illustrates the twelve steps to follow to clone a cow. Dr. Ian Wilmut, pictured here, cloned the first sheep in 1996.**

# FOUR-LEGGED DRUG FACTORIES

Sheep could be genetically altered to produce beneficial drugs in their milk.

Someday soon, medicines might be commonly produced not in laboratories, but in barnyards and meadows! In this excerpt from her book *Clone: The Road to Dolly and the Path Ahead*, science writer **Gina Kolata** explains how cloned sheep could greatly expand the production of useful drugs:

[Such sheep] might produce valuable drugs much more cheaply than did the methods used by drug companies. . . . [The researchers] would clone a lamb whose udder cells made the drug whenever they made milk—all they'd have to do is hook the drug-producing gene to the gene that is turned on when milk is produced and make clones from those genetically altered cells. Then the company could simply milk the sheep, extract the drug from the milk, and sell it. If the scientists made both male and female sheep that carried the added gene, they could breed these sheep and have a self-perpetuating flock of living drug factories.

(above) An entire herd of drug-producing sheep could be cloned from one genetically engineered sheep. (below) A lab technician draws blood from a woman's arm. Many scientists believe that the milk of cloned animals could replace human blood as the source of life-saving drugs.

diseases that impair the normal functioning of the lungs and cause severe shortness of breath.

AAT is usually taken from human blood in a laboratory. The process is costly, though, and there is also a risk because some human blood supplies contain serious diseases, such as hepatitis.

Scientists believed it would be cheaper and safer to "pharm" AAT from herds of cloned animals. (Pharming is a new term that

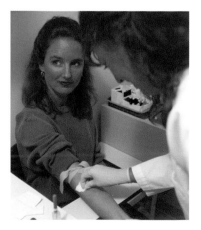

combines the words farming and pharmaceutical. It means "the farming of drugs.") They discovered, though, that genetically engineering sheep to produce AAT in their milk was not an effective way to make large amounts of the drug. One problem is that only a small amount of AAT can be taken from the milk of a single sheep. Also, it is very time-consuming and expensive to

perform the genetic engineering
needed to have even a few
sheep make the drug.

Wilmut and his fellow
researchers saw that
cloning might solve these
problems. They hoped to
clone a sheep that had already
been genetically altered to
produce AAT. That way, all of the animal's offspring would
automatically be programmed to make the drug. These kinds of
cloned sheep would be, in a sense, living drug factories.

Ever since Dolly came into the world, research labs everywhere
have cloned many more sheep. They have also cloned cattle, pigs,
and other animals. Doctors, farmers, and researchers who work for
drug companies watch the ongoing experiments closely. Many
people are confident that animal cloning will benefit humanity in a
number of ways.

**(above) These piglets began life in Dr. Wilmut's genetics lab.
(below) A researcher who worked with Dr. Wilmut to clone Dolly
peers through a microscope.**

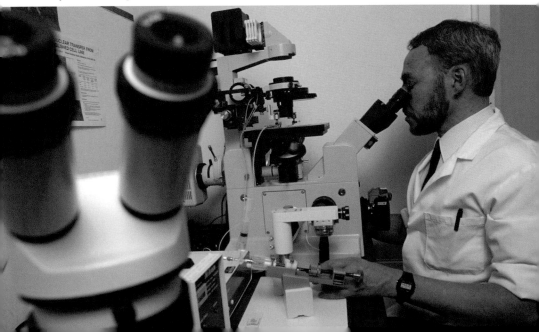

## CHAPTER 2

## THE POTENTIAL BENEFITS OF ANIMAL CLONING

The cloning technology that produced Dolly at the Roslin Institute in 1996 opened up many possible benefits for business and society. Farmers and the food industry, for example, would in theory be able to create whole herds of animals that had been genetically engineered to meet human needs. Eventually, scientists think that companies will also be able to produce tens of thousands of cloned embryos of various types. Each embryo type will have some special quality. For example, one type of cow clone might become an adult that gives more milk than the average cow.

**A rancher's best cow could be cloned into a herd of cattle able to produce great quantities of milk.**

In the future, it is likely that farmers will look at catalogs that list various kinds of animal embryos for sale. The farmers will choose the embryos they want and order them. They will then implant the embryos inside the wombs of their own cows. This simple process could allow a farmer to rapidly create a large herd of cows that all share some trait the farmer wants.

Another goal in the science of cloning is to make cows, sheep, and other animals better able to fight off diseases. Cloning might help farmers produce herds of animals that are more resistant to disease. This medical advantage is only one of many promised by the cloning of animals. Of special interest to numerous scientists and companies are the medical benefits of animal cloning that will be helpful to humans. These include the production of useful drugs in large amounts and for low cost. Another possibility is that animal organs might be transplanted into human patients.

# CURING DISEASE

The ability to make medicines to treat and cure various diseases is one of the most promising potential uses for animal cloning. In fact, the Roslin Institute cloned Dolly in the hopes that larger, less expensive supplies of the drug AAT could be made.

One possible medical use of animal cloning technology is to help hemophiliacs, people whose blood does not clot properly. Hemophilia is caused by a lack of

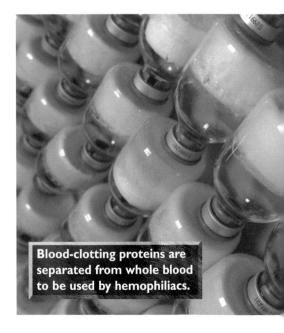

Blood-clotting proteins are separated from whole blood to be used by hemophiliacs.

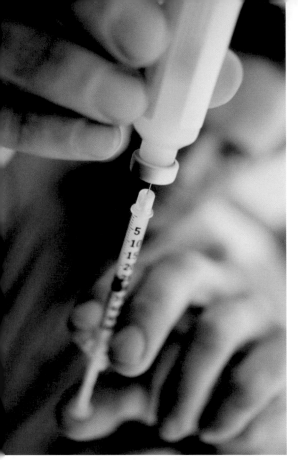

**Someday, insulin may come from cloned animals, which would make the drug less expensive.**

the gene that tells the body to make factors VIII and IX, the proteins that cause blood to clot. Because a hemophiliac's blood does not have one or both of these factors, he (most hemophiliacs are male) can suffer significant internal bleeding from only a minor injury. It may be possible to genetically alter certain animals so that the milk they produce will contain the missing clotting factors. It would then be fairly easy for lab technicians to separate the clotting factors from the milk and make a drug to treat hemophilia.

This same approach may also be used to make insulin. Insulin is a hormone that is made in the body and is essential to breaking down carbohydrates. Diabetes, which affects one in 20 people in the United States, is a condition in which a person's body does not produce enough insulin. As a result, a diabetic person often has unhealthy levels of sugar and other substances in the bloodstream. To avoid serious illness, many people who suffer from diabetes must take insulin daily. Several new companies have made it their goal to genetically alter cows or sheep to produce insulin, then clone the animals to create large herds capable of creating large amounts of the drug. That way, it would be both faster and cheaper to make insulin than it is today, which would help make the drug more affordable.

# MORE EFFICIENT RESEARCH ANIMALS

Animal cloning technology will greatly help scientists improve the quality of laboratory experiments. For almost two centuries, lab researchers have used mice, rats, guinea pigs, and other small animals to study how various diseases affect living things. Lab animals are also used to test new medicines designed to fight diseases. Researchers have always had to deal with the fact that their lab animals were not identical. Just as humans differ from one another both physically and mentally, so do animals, but it is best to have all the animals in a particular test be as much alike as possible. The traditional way to achieve this was the long process of selective breeding. White rats of a certain size would be mated with other white rats of that size. This allowed lab technicians to produce many white rats of about the same size and weight. Despite appearances, however, these rats were still genetically different. In a drug test that used the animals, a researcher could not be certain if each rat's reactions to the drug were caused mainly by the drug or by unknown differences in the rats themselves. In contrast, research animals bred through cloning will be genetically identical. That means the results of most experiments will be much more accurate.

# ANIMAL ORGANS FOR HUMAN BODIES

Animal cloning also shows great promise for transplants of animal organs and parts into people. Among the possibilities are transplants of hearts, kidneys, and bone marrow. Human donor organs are always in limited supply. In fact, about 3,000 people in the United States die each year as they wait for the organ transplants they need to save their lives. This problem might be partially solved by cloning animals and harvesting their organs.

Although scientists have experimented with transplanting animal organs into humans, one major drawback of such transplants is that the recipients' bodies usually reject the animal

**Surgeons may one day be able to transplant organs harvested from clones of genetically engineered pigs.**

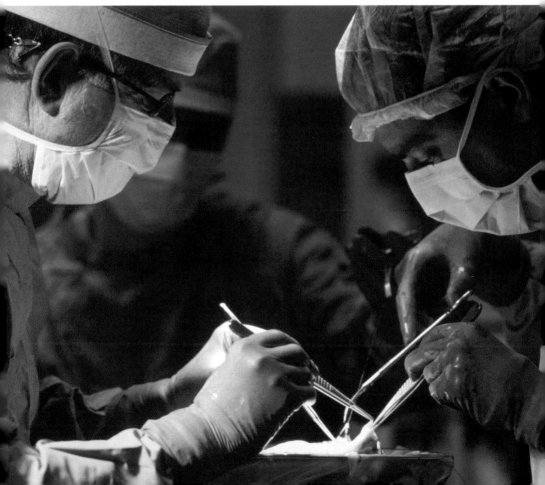

**Scientists believe that pigs' organs can successfully be transplanted into humans.**

organs. In fact, it is the human body's natural response to reject foreign organs. Even organs that come from other humans are often rejected.

To solve this problem, scientists seek to change animals to make their organs a better match for human bodies. Put simply, scientists will add certain human genes to the animal's own genes. The goal is to reduce the chance that the human body will reject the animal's organs. At present, most researchers think that pigs show the most potential for this kind of procedure. There are two reasons for this. First, researchers say, humans are less likely to contract diseases from pigs than from monkeys and other primates. (Because humans are primates, they could more easily catch diseases from other primates.) Second, the size and shape of a pig's organs are very similar to a human's.

# SAVED FROM THE BRINK OF EXTINCTION

Yet another benefit of animal cloning is the ability to save endangered species. When the Roslin Institute cloned Dolly, some scientists saw that it was possible to clone an existing member of an animal species to create new, healthy members of that species. More important, they realized, this technique should work just as well even if the animal in question is the very last of its kind.

Not long after Dolly's creation, the process of cloning did save an animal from extinction. The creature was a rare breed of cow— the Enderby—that is native to New Zealand. By the mid-1990s, only one Enderby cow was left. In 1996, some New Zealand scientists decided to try to clone the cow, named Lady, to see if the technique that had created Dolly might also be used to save an endangered species. They grew some embryos from cells taken from Lady's ovaries. Next, they placed the embryos in the wombs of some cows of a different breed, which played the role of surrogate mothers. In July 1998, the first new Enderby calf came into the world. This success showed that endangered species could, in fact, be rescued through cloning.

Researchers foresee that cloning technology will enable zoos to save endangered animals that do not breed well in captivity. The giant panda is probably the best-known example. Only a few hundred of these animals exist in the world today. To save the pandas, a group of Chinese scientists is now attempting to clone some of them. The program shows a lot of promise.

# BRINGING BACK DINOSAURS?

Some scientists think it might be possible to go even further and resurrect species that are already extinct. They point out that the genetic material needed to create clones does not have to come from a living creature. In theory, as long as the DNA is intact, the

The endangered giant panda is a prime candidate for cloning.

# BRINGING ROVER BACK OVER

Ever wish you could bring back a beloved pet that died? If so, you are not alone. A number of people have expressed an interest in cloning their dead pets. There would be certain limitations and drawbacks to this

People may some day clone their favorite pets.

kind of procedure, though. First, although the cloned animal would look just like the original pet, it would not have the same memories, exact intelligence level, or personality. Therefore, the owner would not really be able to resurrect all the qualities of the pet that he or she once knew and loved.

Another problem is the high cost of cloning. Such a procedure would cost many thousands of dollars, beyond the means of most pet owners. Of course, a few people would be able to afford it. Those people will have several companies to choose from, some of which have already begun to prepare for the pet-cloning market. One of these companies is Geneti-Pet of Washington State, which believes that cloning pets will soon be both possible and profitable.

preserved cells of an extinct animal might be cloned to produce a living copy of that animal.

If this is the case, might scientists be able to bring dinosaurs back to life, as is done in the popular book and movie *Jurassic Park*? One major obstacle stands in the way of this achievement. To clone a dinosaur, scientists would need to find a substantial amount of its DNA in usable condition. The problem is that dinosaur remains are hard to find and are badly decayed. This

**Scientists have been unable to find enough usable DNA to clone a dinosaur.**

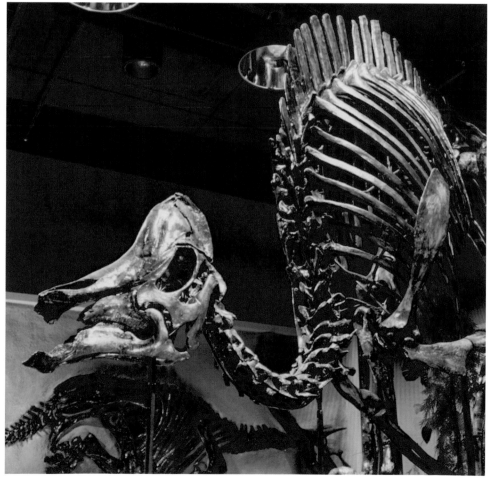

makes the chances of finding any usable DNA slim at best. After animals die, their bodies begin to break down. Nature uses these materials to make new soil, plants, and animals. These, in turn, die and leave their own remains for the next generation. Among these recycled materials is DNA, which can no longer be used once it has broken down.

In *Jurassic Park*, the scientists find a way to get around this problem. They discover some dinosaur DNA in blood in the stomachs of dinosaur-biting insects. The insects' bodies have been preserved for millions of years in amber—hardened tree sap. It is true that scientists have found pieces of amber that contain the DNA of insects and plants about 30 million years old. Scientists have not been able to collect enough of any ancient creature's genetic material, however, to make cloning it possible. Therefore, the chance that enough intact DNA to clone a dinosaur might be found is remote. Only if some presently unknown way is found to get dinosaur DNA will the cloning of dinosaurs become a real possibility.

## CHAPTER 3

# THE PROMISE OF HUMAN CLONING

The successful cloning of Dolly the sheep and other animals in the 1990s naturally raised the question of whether human beings might also be cloned. If sheep and cows could be cloned, many people said, it might be just a small leap to the cloning of humans. People also wondered whether it would be ethically or morally right to clone humans. Many religious leaders and politicians quickly answered this question in the negative. They argued that to clone humans would be, in effect, to tamper in God's domain.

**Many people oppose cloning for religious and ethical reasons.**

Those who disagree with this position argue that large numbers of people of all religious backgrounds support the idea of cloning research. Moreover, they say, no single religious group should be able to force its views on everyone else. Furthermore, the failure to develop and use human cloning might harm society in the long run if it holds up medical breakthroughs and other benefits. Most scientists say that no matter what people may believe about the morality of human cloning, the technology will become a reality sooner or later. The real question, they say, concerns which nations or groups will take steps to regulate it and benefit from its vast potential.

**This animal trainer cloned his pet bull that had died years earlier. Many want to use cloning to bring back deceased pets.**

# BRINGING BACK THE DEAD

Scientists point out that many objections to human cloning are based on incorrect ideas of what it really is and how it will be used. Those who oppose the idea of cloning people often cite what they see as unhealthy, even twisted, motives. Critics used to warn that dictators might create armies of mindless followers or evil henchmen, or that unethical businesspeople would grow cloned bodies to produce human organs for sale.

Plans to use cloning for global domination have been shown to be unrealistic and are no longer taken seriously. To clone an army of followers would take decades and cost hundreds of billions of dollars. It would be very difficult to keep such a project secret. Moreover, although the cloned followers might be physically identical, they would all be individuals with separate minds and wills. It would be impossible to make them all think and act the same. As for cloning bodies for organs, it would be far cheaper and easier to harvest the organs of cloned animals or to grow individual human organs in a lab.

For many ordinary people who support cloning, a much more realistic aim is to bring back deceased loved ones. Thousands of people have expressed the desire to resurrect beloved pets, and several companies have already started to prepare to meet this demand, which is expected to grow in the future. In a similar vein, many people say they would like to have clones of children who have died. A number of medical clinics, newspapers, and Internet sites receive hundreds of inquiries each month that ask how loved ones might be brought back through cloning.

At least some of these curious people are less sure about human cloning once they learn how the process really works. Many people still mistakenly believe that the clone of a dead person will be an exact replica. The reality is quite different. The act of cloning a person (or pet) will duplicate only his or her physical appearance.

This human embryo was created in a laboratory with cloning techniques. Some people hope cloning could be used to resurrect loved ones who have died.

Personality and memories cannot be resurrected, so the clone's feelings, desires, and opinions will be different from those of the deceased. (Even after they learn this, a large number of people who hope to bring back deceased loved ones say they still want to do it.)

# ARMIES FROM THE LAB?

Budding dictators who hope that cloning will offer a way for them to create ready-made armies, subject to their every whim, are destined to be disappointed. In this excerpt from his book on cloning, writer Daniel Cohen explains why the once common idea that human cloning

These women want to be surrogate mothers for clones. Vast numbers of such women would be needed to produce clones on a large scale.

might lead to the creation of evil armies is not feasible:

One of our darkest fears about cloning is the . . . image of hordes of clones produced for specific purposes, usually menial labor or warfare. Such fears come largely out of the misconception that cloned humans or sheep are actually "grown" in the laboratory. We have all heard the phrase "test tube baby." But that's not the way [cloning] works. . . . To produce large numbers of human clones, given the high failure rate in cloning procedures, [very large] numbers of women would have to agree to undergo the repeated and sometimes risky surgery necessary to obtain eggs. . . . Even more significantly, vast numbers of human surrogate mothers would be required. The number of women who would be willing essentially to rent out their wombs in order to grow clones is never likely to be very large. And the expense of producing even a single human clone would be enormous. It is far easier, safer, cheaper, and more pleasurable to reproduce offspring the old-fashioned way. That is not going to change. . . . The frightening image of mass production of laboratory-grown clones is now strictly science fiction.

# HELPING CHILDLESS COUPLES

Experts on cloning say that, if it happens at all, the cloning of deceased loved ones will likely account for only a small portion of the future uses of human cloning. Indeed, it is more likely that cloning technology will help people in less controversial ways. One way would be to give infertile couples a way to have a child who is genetically related to them.

Today, there are two main ways for infertile couples to have a biological child. The first is in vitro fertilization, in which lab technicians join a man's sperm and a woman's egg in a lab and implant the resulting embryo into the woman's uterus. The major drawback of this approach is that it does not work for everyone. A second method can be used in cases where a couple's sperm or eggs are absent or not viable. For instance, an infertile man and his wife

**A scientist performs an in vitro fertilization in a laboratory. In vitro fertilization is an effective reproductive technology, but it does not work for everyone.**

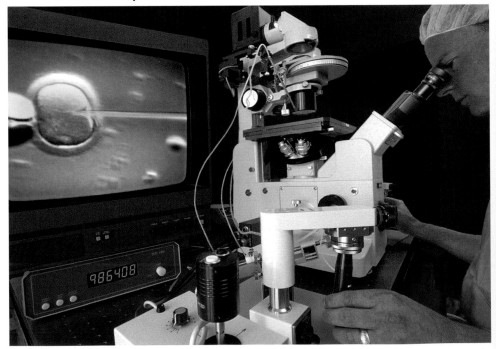

Even infertile couples for whom other reproductive technologies have failed could have a child through cloning.

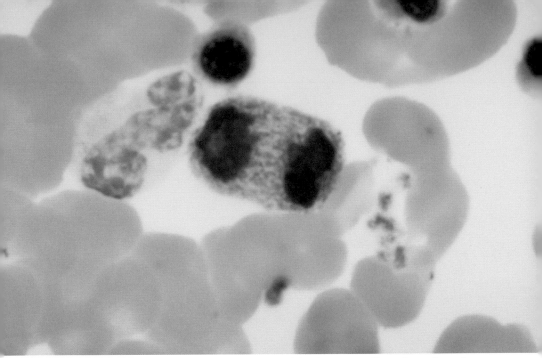

Genetic material taken from any healthy human cell, such as these bone marrow cells, could be used to produce a baby. Other reproductive technologies rely on sperm and egg cells, which are not always viable.

might use sperm donated by another man. This sperm is combined with the woman's egg in the lab to make an embryo. The problem with this approach is that the baby will be genetically related to only one of the parents.

This type of problem will be eliminated once human cloning is perfected. Even if a man were infertile, genetic material could be taken from some of his cells and transplanted into the woman's egg to produce offspring. Such techniques may also be able to help another group of people who have traditionally been unable to have biological children—people with serious genetic disorders. Normally, if either parent suffers from such a disorder, they worry that the defective gene will be passed on to their children. One potential solution to this problem is to clone a baby from a cell taken from the parent who does not have the disorder. Although the child would be genetically related to only one of the parents, many couples say they find this idea appealing.

# NEW TISSUES AND ORGANS

A number of researchers say that childless couples are not the only people who will put human cloning technology to use. Cloning could also be used to provide replacement organs and tissues for sick or injured people. Patients in need would place orders for organs, and technicians would grow lungs, kidneys, or other organs in labs. Unlike the cloning of organs from pigs and other animals, a technique in which the required organs or tissues could be made from the patient's own cells would eliminate the risk of rejection, since there would be an exact genetic match.

Effective methods that would allow labs to grow human organs and tissues were pioneered by American and English scientists beginning in 1981. These researchers wanted to learn to control the process by which the body's cells differentiate, or become specialized. Normally, when embryos first form, their cells all look

**Human skin tissue is made up of specialized cells (pictured). Scientists, however, can grow skin and other tissue from nonspecialized stem cells.**

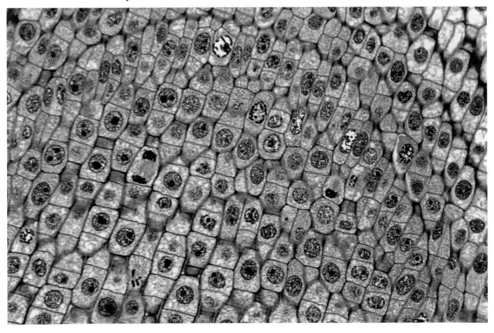

and act more or less the same. Eventually, though, the cells begin to differentiate—some become heart cells, others become skin cells, and still others become bone marrow cells, for instance. These and other specialized kinds of cells combine to make up a human fetus, which later becomes a baby.

In a significant breakthrough, researchers were able to stop the process of differentiation in a mass of embryonic cells. The cells continued to reproduce themselves, but all of the new cells that grew were the simple, nonspecialized embryonic type. Researchers call these cells embryonic stem cells, or ES cells. ES cells have enormous potential. Scientists believe they will be able to send certain signals that will fool the cells and keep them in an embryonic state. Other signals will tell the cells to grow into specific kinds, such as blood cells or nerve cells.

**An embryonic stem cell (pictured) can become any type of cell. Cloning combined with stem cell technology could have many medical benefits.**

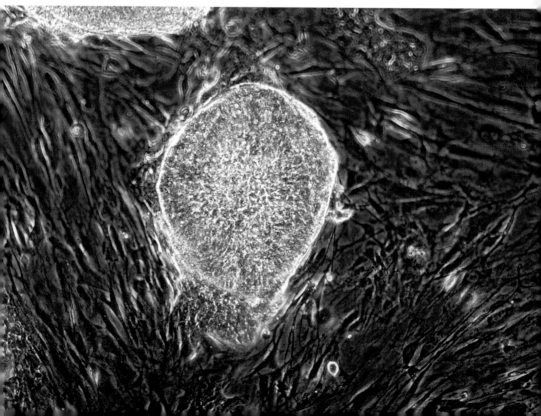

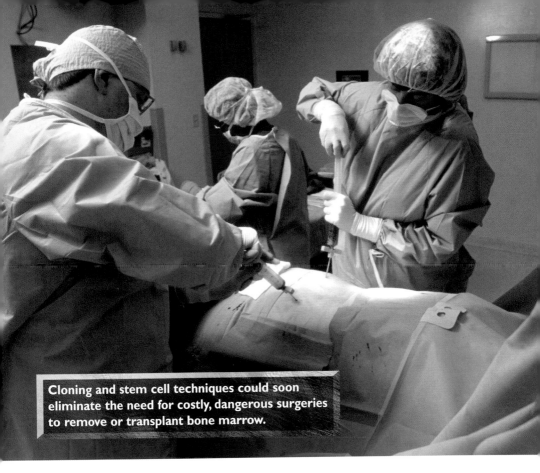

Cloning and stem cell techniques could soon eliminate the need for costly, dangerous surgeries to remove or transplant bone marrow.

The benefits that might result from the combination of cloning and ES cell technology are great. The new technology could, for example, provide a cure for a patient who has leukemia (a blood disease that is usually fatal). Today, a bone marrow transplant can sometimes save a leukemia patient's life. There must be a close genetic match between the donor and the patient, however, and it is often difficult to find such a match. There is always the risk that the patient's body will reject the donated marrow.

Cloning would be a better solution. Doctors could take a cell from a leukemia patient and clone it to make an embryo. Then the doctors would apply certain chemicals to the embryo that would signal the embryonic cells to grow into a mass of ES cells. Another chemical signal would order the ES cells to become bone marrow cells. Eventually, the doctors would have enough new bone

**This child's damaged spinal cord may someday be repaired using cloning techniques.**

marrow cells to inject into the patient. The patient's body would not reject these new cells because they would carry the patient's own genetic code. The leukemia would be cured, and the patient would have a chance for a long and healthy life.

Such life-saving transplants are not the only medical benefits that may come from the combination of ES cell and cloning technologies. Burn victims will be given new, healthy skin. People who suffer from brain and spinal cord damage will receive new brain and spinal cord cells grown especially for them. Vital organs, such as hearts and lungs, will be grown in labs, too. Finally, the ability to alter and redirect the growth of cells may well lead to cures for most or even all types of cancer.

These possible uses of cloning are a long way off, though. Cloning research is still in the very early stages. Many technical problems have to be overcome before human cloning will be workable and safe. Even so, a growing number of people believe that when that day comes, humanity will enjoy many amazing medical advances, as well as other benefits that no one yet foresees.

# GLOSSARY

**DNA** The major component of the genetic material of living things.

**embryonic stem cells (ES cells)** In an animal or human embryo, cells that have not yet differentiated, or begun to develop into specific kinds of cells, such as nerve cells or bone marrow cells.

**extinction** The death of a species.

**genetic disorder** A hereditary disease caused by a defective gene passed from parent to child.

**genetics** The study of heredity.

**heredity** The transmission of physical characteristics from one generation to another.

**pharming** A new word that means "the farming of drugs."

# FOR FURTHER INFORMATION

*Books:*
Daniel Cohen, *Cloning.* Brookfield, CT: Millbrook Press, 1998.
*A very well written, easy to understand, and interesting exploration of cloning and modern conceptions of it.*

Michael Crichton, *Jurassic Park.* New York: Knopf, 1990.
*The famous novel that inspired the equally famous film about scientists who successfully clone dinosaurs from genetic material found in the digestive tracts of ancient insects preserved in amber. Although it is fiction, it raises and explores questions about the cloning of extinct animal species that scientists have actually begun to address.*

Rob DeSalle and David Lindley, *The Science of* Jurassic Park *and the* Lost World, *or How to Build a Dinosaur.* New York: Basic Books, 1997.
*A brilliant and fascinating look at the scientific technology and possibilities involved in the cloning of extinct animal species. Highly recommended for ambitious young readers.*

Jeanne DuPrau, *Cloning.* San Diego: Lucent Books, 2000.
*A well-organized, thorough, and informative overview of the cloning phenomenon, covering plant cloning, animal cloning, possible medical benefits of cloning, and the ethical and legal considerations of possible human cloning.*

David Jefferis, *Cloning: Frontiers of Genetic Engineering.* New York: Crabtree, 1999.
*Beautifully illustrated with color photos and drawings, this is an excellent introduction for young people to cloning and related topics, such as genes, cells, and DNA.*

Gregory E. Pence, ed., *Flesh of My Flesh: The Ethics of Cloning*. New York: Rowman and Littlefield, 1998.
*Discusses the various ethical and moral arguments for and against cloning.*

Jon Turney, *Frankenstein's Footsteps: Science, Genetics, and Popular Culture*. New Haven, CT: Yale University Press, 1998.
*A brisk exploration of changing public attitudes about cloning and the way the public often fears or sees a distorted image of science and its newer, more controversial creations.*

Lisa Yount, ed., *Cloning*. San Diego: Greenhaven Press, 2000.
*A very well chosen collection of essays about cloning, it covers the cloning of Dolly and its importance, the cloning of endangered species, movies and other media presentations of cloning, legal implications of cloning, ethical concerns, and much more. The reading level is high school, but not too challenging for those younger readers who have a strong interest in cloning.*

*Articles:*
Sharon Begley, "Little Lamb, Who Made Thee?" *Newsweek*, March 10, 1997.

Jon Cohen, "Can Cloning Help Save Beleaguered Species?" *Science*, May 30, 1997.

J. B. Gurdon and Alan Colman, "The Future of Cloning," *Nature*, December 16, 1999.

Gina Kolata, "Researchers Find Big Risk of Defect in Cloning Animals," *New York Times*, March 25, 2001.

Michael Mautner, "Will Cloning End Human Evolution?" *Futurist*, November/December 1997.

Charles Pellegrino, "Resurrecting Dinosaurs," *Omni*, Fall 1995.

Richard Stone, "Cloning the Woolly Mammoth," *Discover*, April 1999.

Alan Taylor, "Silence of the Lamb," *New Yorker*, March 17, 1997.

*Websites*
*Each of the following sites provides several links to up-to-date, reputable, and factual information about cloning and the many scientific, social, and ethical issues that surround it.*

Cloning Fact Sheet of the Human Genome Project:
http://www.ornl.gov/hgmis/elsi/cloning.html

New Scientist.com: http://www.newscientist.com/hottopics/cloning

Roslin Institute Online: http://www.ri.bbsrc.ac.uk/library/research/cloning/cloning.html

Yahoo News: http://dailynews.yahoo.com/fc/Science/Cloning

# ABOUT THE AUTHOR

In addition to his acclaimed volumes on ancient civilizations, historian Don Nardo has published several studies of modern scientific discoveries and phenomena. Among these are *The Extinction of the Dinosaurs, Atoms, Vaccines, Black Holes, and The Solar System.* Mr. Nardo lives with his wife, Christine, in Massachusetts.

# INDEX

AAT, 18, 20-21, 23
Agriculture, 5, 8, 14–15, 22
Alpha-1 antitrypsin (AAT), 18, 20-21, 23
Animals, 5, 8, 10, 12, 16–21, 22–32, 33
 organs for humans, 26–27, 41-44
 research, 25
Apples, 14–15
Asexual reproduction, 7

Bacteria, 8, 12
Blue-green algae, 12
Bone marrow transplant, 43-44
Books, 4-5
*Boys from Brazil, The*, 5, 11

Cancer, 44
Cells, 7, 17, 40, 41, 42, 44
Childless couples, 38, 40, 41
Cloning
 in agriculture, 14–15
 and animals, 5, 8, 10, 12, 16–21, 22–32, 33
 costs, 30
 defined, 7
 disadvantages, 15
 and food, 8, 15, 22
 human, 33–44
 in nature, 12
 and plants, 5, 8, 10, 12, 14, 16
Cohen, Daniel, 37

Cows, 28, 33
Cystic fibrosis, 18

Death, 35–36, 38
Diabetes, 24
Dinosaurs, 31–32
Diseases, 5, 15, 18, 25, 27, 43–44
 curing, 5, 23–24
Diversity, 15
DNA, 7, 14, 31, 32
Dolly, 18, 19, 21, 22, 23, 28, 33
Drugs, 18, 19, 20-21, 23–24, 25

Eggs, 17, 18, 38
Embryo, 17, 23, 28, 38, 40, 41–42
Embryonic stem cells, 42, 43
Emphysema, 18
Endangered species, 28
Ethics, 4, 33
Experiments, lab, 25
Extinction, 28, 31

Fetus, 42
Food, 8, 15, 22
Frogs, 18

Genes, 7, 14, 27
Genetic disorders, 40
Grapes, 14

Hemophilia, 23–24
Hitler, Adolf, 5, 11
Humans
    and animal organs, 26–27, 43
    and cloning, 33–44

In vitro fertilization, 38
Infertility, 38, 40
Insulin, 24

*Jurassic Park*, 31, 32

Laboratory experiments, 25
Leukemia, 43–44
Levin, Ira, 5, 11
Livestock, 10, 16

Medicine. *See* Drugs
Memories, 36
Morality, 33, 34
Mothers, surrogate, 37
Movies, 4–5

Nature, 12
Nucleus, 17, 18

One-celled organisms, 8, 12
Organ transplantation, 23, 26–27, 43
Organs, replacement, 41–44

Pandas, 28
Personality, 11, 36
Pets, 30, 35
Pharming, 20
Pigs, 27, 41
Plants, 5, 8, 10, 12, 14, 16
Protozoa, 12

Replacement organs, 41–44
Reproduction, 5, 7, 8, 12, 37
Research animals, 25
Roslin Institute, 18, 22, 23, 28

Selective breeding, 25
Sexual reproduction, 7
Sheep, 18, 19, 20-21, 33
*Sleeper*, 4
Spemann, Hans, 17
Stem cells, embryonic, 42, 43-44
Surrogate mothers, 37

Tissues, replacement, 41–44
Transplants, organ, 23, 26–27, 43

Water hyacinth, 12
Wilmut, Ian, 18, 21

Yeast, 12